Taiane Santos Marques
Gerlany F. S. Pereira

Transgenics in Basic Education

Taiane Santos Marques
Gerlany F. S. Pereira

Transgenics in Basic Education

Knowledge of students from public schools in Santana-AP in focus

ScienciaScripts

Imprint
Any brand names and product names mentioned in this book are subject to trademark, brand or patent protection and are trademarks or registered trademarks of their respective holders. The use of brand names, product names, common names, trade names, product descriptions etc. even without a particular marking in this work is in no way to be construed to mean that such names may be regarded as unrestricted in respect of trademark and brand protection legislation and could thus be used by anyone.

Cover image: Provided by the author

This book is a translation from the original published under ISBN 978-620-2-40789-2.

Publisher:
Sciencia Scripts
is a trademark of
Dodo Books Indian Ocean Ltd. and OmniScriptum S.R.L publishing group

120 High Road, East Finchley, London, N2 9ED, United Kingdom
Str. Armeneasca 28/1, office 1, Chisinau MD-2012, Republic of Moldova, Europe
Printed at: see last page
ISBN: 978-620-8-32489-6

"Let your efforts defy impossibilities, remember that the great things of man have been conquered from what seemed impossible.

(Charles Chaplin)

DEDICATION

I dedicate this book to my parents, husband, daughter, family and friends who have encouraged and helped me in so many ways to make it possible.

(TAIANE SANTOS MARQUES)

To science teachers all over Brazil, who work hard to build a citizen's education!

(GERLANY DE FATIMA DOS SANTOS PEREIRA)

ACKNOWLEDGMENTS

First of all, I thank God for allowing me to experience this moment.

To all the professors of the Natural Sciences course at Amapa State University, who have been so important in my academic life, especially my advisor.

(TAIANE SANTOS MARQUES)

Thanks to everyone who made this research possible!

(GERLANY DE FATIMA DOS SANTOS PEREIRA)

SUMMARY

The aim of this study was to investigate the knowledge of 3rd year high school students from two public schools in the municipality of Santana-AP about transgenic foods (TA). This is a qualitative-quantitative study in which 60 third-year high school students from two public schools in Santana took part. To collect the data, a questionnaire was used, containing open questions aimed at investigating the students' knowledge and interest in transgenics; what they understood by biotechnology and the advantages and disadvantages of consuming TA. The Discourse of the Collective Subject was used to analyze the data. The answers were categorized and then discussed with the literature based on the inferences made by the researcher. The results showed that in the schools investigated, the teaching of biotechnology and, consequently, of TAs is still developed in a mechanized way by high school teachers, which results in superficial or, in some cases, non-existent knowledge on the part of the students. It can therefore be deduced that the results found may be due to various factors, ranging from inadequate teacher training (initial and continuing) to approach the subject in a way that is meaningful to the student, to these students' lack of access to practical classes and differentiated teaching methodologies that allow them to assimilate knowledge in a meaningful way.

Key words: Biotechnology. Genetically modified organisms
Modified. Science Teaching and Learning. Scientific Literacy.

INDICE

PRESENTATION

This research addresses the issue of Transgenic Foods (TA) in high school. This is a current and relevant issue in the social context experienced by a large portion of the world's population (PEREIRA, 2012; PEREIRA; RIBEIRO; FREITAS, 2014; PEREIRA; RIBEIRO; FREITAS, 2016). Since TAs are part of the menu of thousands of people around the world, many of them end up consuming them without even knowing it (PEREIRA, 2012; PEREIRA; RIBEIRO; FREITAS, 2014; PEREIRA; RIBEIRO; FREITAS, 2016).

Thus, the interest in investigating the subject arose from observing the major objective of science teaching. It is the Scientific Literacy (SL) of its citizens, which requires them, in Chassot's (2014) terms, to be able to read and understand science, participating actively and consciously in the society in which they live. Therefore, a subject as relevant as TA should be something that students have mastered by the end of their basic education, i.e. the 3rd year of secondary school.

With this in mind, the following research question arises: What knowledge do 3rd year high school students from public schools in Santana-AP have about biotechnology and TA?

Thus, as a hypothesis, it is believed that students' knowledge of biotechnology and TA at the end of primary school is superficial or insufficient to enable them to make decisions based on scientific knowledge. As this is a subject that requires a level of initial and ongoing training, it demands constant updating on the part of teachers. This training is not always available to teachers who work in science education.

The aim of this research was to investigate the knowledge of 3rd year high school students from two public schools in the municipality of Santana-AP about TAs.

1 INTRODUCTION

The importance of discussing aspects related to teaching about transgenic foods (TA) stems from the fact that this is a much-discussed subject today (PEREIRA, 2012; PEREIRA; RIBEIRO; FREITAS, 2014; PEREIRA; RIBEIRO; FREITAS, 2016). However, it is not new, given that since the end of the 1970s, when researchers discovered how to transfer genes from one organism to another, whether animal or plant, altering their natural characteristics, thousands of people have become interested both in discovering the benefits that the "new" practice could bring to man, and in possible problems arising from it.

However, there are still schools that are not always able to keep up with developments on the subject, as some teachers only use textbooks as a methodological resource, which ends up hindering the student's teaching and learning process. By studying biotechnology and its aspects, and in view of the problems raised, I became interested in investigating the knowledge of high school students on the subject of TA. Since, at the end of this stage of schooling, these students need to be able to make informed decisions, based on scientific knowledge, and should be scientifically literate. The following guiding question was therefore drawn up: what knowledge do 3rd year high school students from public schools in Santana-AP have about biotechnology and TA?

These approaches are justified by the fact that transgenic food products are offered to consumers in specific stores and large supermarket chains. However, because people have little knowledge about them, they prefer not to use them. In this context, it is important that the subject is addressed more emphatically in the school environment, specifically in high school, so that students acquire more knowledge about the subject, so that they can question, analyze and decide whether or not to consume GMOs.

It was after observing this need that it was decided to carry out a survey in public high schools in the city of Santana, Amapa state, to find out whether teachers use new methodologies, or the help of the media, among others, to address the issue, helping students to understand the subject and be able to discuss it both in the classroom and in their social context, their community.

The general aim of this research was to investigate the knowledge of 3rd year high school students from two public schools in the municipality of Santana-AP about TAs. To this end, a qualitative-quantitative study was carried out, using a questionnaire as the data collection technique.

2 BIOTECHNOLOGY

Takahashi; Martins and Quadros (2008) state that biotechnology is an increasingly broad term, involving the development of techniques, products or processes using living organisms or parts of them. The authors consider biotechnology to be a multidisciplinary area that is developing, encompassing various areas of knowledge such as Genetics, Biochemistry, Agronomy, Microbiology, Pharmacology, Chemistry, Immunology, among others.

For Borem and Carneiro (2008), biotechnologies, in their broadest sense, include the manipulation of microorganisms, plants and animals, with the aim of obtaining processes and products of interest. In this way, according to the authors, any activity that involves the application of knowledge of physiology, biochemistry and genetics is considered to be biotechnological. In its strictest sense, biotechnology is associated with the use of modern molecular and cellular biology techniques.

When discussing biotechnology, Gongalves (2010) argues that it has been used since ancient times in the production of bread and fermented drinks, but in a very artisanal process. Nowadays, according to the author, biotechnology uses state-of-the-art techniques and materials, which has led to a huge expansion in its field of application.

With the emergence of studies in microbiology (fermentation of beverages) and molecular biology (tissue culture), knowledge of the manipulation of microorganisms and genes has made it possible to produce various medicines and industrialized foods, according to Gongalves (2010). Examples of biotechnological advances include insulin, which is produced by genetically modified bacteria, and the production of medicines using monoclonal antibodies.

Furthermore, as Durbano et al. (2008) point out, biotechnology has never been discussed as much as it has been in recent decades, especially when we think of it as a source of resources that help to improve people's lives. The authors cite as examples the advances in science, humanity's growing demand for food, cures for epidemics, among others.

Durbano et al. (2008) also point out that there are a number of issues involving the application of biotechnology that affect the lives of citizens, such as DNA-based diagnostics, cloning, stem cells and transgenics. The authors also argue that the population has a personal opinion on these issues, making it clear that there is a need to promote concepts relating to this new science, in order to ensure that these technologies are used for the benefit of all.

With regard to the benefits that biotechnology can bring to mankind, these are mainly reflected in the areas of health and food production. In this respect, it is important to consider what Gongalves (2010) says when he states that much of what we eat and use as medicines is the work of biotechnology.

Gongalves (2010) also points out that among the various uses of biotechnology, there are the benefits it provides to agriculture, where it is used on a large scale in the production of transgenic organisms, i.e. the addition of a gene that codes for a characteristic of interest to the genome of another plant. This gene can be from a fungus, a bacterium or even another plant.

Pereira; Ribeiro and Freitas (2016) consider that biotechnology teaching will only have satisfactory results if the methodologies and resources, as well as the content, are socialized knowledge that makes sense to the student. In other words, the student must understand the importance and use of what they learn in their day-to-day lives, making use of it in their social environment and that it serves to modify the social context in which they are inserted, so that learning is actually taking place, consolidating itself as expected by the teacher, the school and society.

3 TRANSGENICS

According to Nodari and Guerra (2000), transgenics is a technique that can make a significant contribution to the genetic improvement of plants, with a view to producing food, fibers and oils, as well as pharmaceuticals and other industrial products. The authors warn, however, that there are a number of challenges to overcome in order to reap the benefits of using modern biotechnologies.

For some authors, such as Furnival and Pinheiro (2008), it was precisely the controversies that arose at the end of the 20th century around genetically modified organisms (GMOs) - the so-called transgenics - that led to society's increased interest in technological innovation processes in the biotechnology area.

Discussions aside, according to Lima (2005), TAs or GMOs are revolutionizing food culture. According to the author, they differ from other foods mainly because they have a high genetic load from another being, and the beings involved in the process do not need to be similar or related for the transgenic process to be possible.

Silva (2005) defines them as organisms whose genetic structure has been altered by the activity of genetic engineering, which uses genes from other organisms to give them new characteristics. This alteration, according to the author, can be aimed at improving the nutrition of a food or making a plant more resistant to a herbicide.

In order to emphasize the importance of transgenics, Lima (2005) points out that they represent a great hope for humanity, given how and why they "came about". However, the author states that it is known that the impact of this discovery has brought a certain contradictory fear of GMOs in general, pointing out that there is a comparison in social behavior in relation to transgenics with cloning, as both come up against ethical issues; scientific, social, ethical and religious.

According to Lima (2005), there are researchers who claim that TAs have already been tested enough to be exposed to the public, while there are others who claim that more caution should be exercised when dealing with this issue, as there is no convincing proof of the veracity of the benefits or harms of these GMOs, generating even more controversy[1] as Weid (2005) and Gorgen (2000) point out.

Lima (2005) also points out that allied to this conception are arguments based on Darwin's theory, which states that living beings, as they are today, are the result of a slow and gradual process of Natural Selection (DARWIN, 1859). In this way, the introduction of GMOs into the environment is unpredictable in the short term, without them being carefully studied.

Lourengo and Reis (2013) also consider such discussions when they state that people have formed their opinions about transgenics mostly through journalistic texts, which either praise them with often absurd hopes or abhor them with unfounded fears, and it is rare that the information is given with due scientific accuracy.

According to Pereira (2012), every day we are "bombarded" with a large amount of information disseminated by the most diverse media and this information generally refers to facts whose understanding depends on mastering scientific knowledge.

Issues such as decision-making and the formation of moral reasoning, for example, are highly emphasized in some of these studies, and barely mentioned in others. But in general, there is agreement that

[1] On this subject, see Pereira (2012). The researcher takes a broad approach to the topic in her text "Transgenic foods: the polemics that permeate debates and discussions on the subject", in her master's research entitled "Appropriation of scientific knowledge: an approach to transgenic foods".

these controversial issues divide the opinions of scholars and various segments of society (PEREIRA; RIBEIRO; FREITAS, 2016).

When addressing issues inherent in the discussions generated by transgenics, Gorgen (2000) takes a brief look back at events and clarifies that until a few years ago this was a technique that caused reluctance among many. Then, when the technique made it possible to produce insulin, growth hormone and *interferon* through transgenic bacteria, there was no expression of doubt or disagreement with their application to human beings. This acceptance came from the fact that the products are of great relevance to medicine and before the technique they were difficult and much more expensive to produce (GORGEN, 2000).

4 TEACHING SCIENCE IN SECONDARY SCHOOL

The literature shows that Genetics has played an important and crucial role in Biology[2] and, due to its various manifestations, modern society maintains a close relationship with this area of knowledge. The author points out that understanding the basic principles of Genetics is fundamental if students are to be prepared to give an informed opinion on the innovations introduced by science into society, thus helping them to exercise their citizenship.

In this sense, Arruda; Branquinho and Bueno (2006) consider that teachers should take advantage of students' natural curiosity to awaken their interest in science, encouraging them to question what they hear or read in the media. The authors warn, however, that we still have to consider the importance of science in various other issues, which are not always in the news, but which are fundamental to the formation of the individual.

According to Brasil (2006), the teaching of biology in secondary school must face a number of challenges, one of which is to enable students to take part in contemporary debates that require biological knowledge. Arruda; Branquinho and Bueno (2006) argue that it is necessary to make students understand that scientific knowledge is the result of a long historical process, which includes mistakes and successes and produces provisional truths.

Arruda; Branquinho and Bueno (2006) point out that this will enable students to realize that the products generated by scientific knowledge are the result of a combination of nature and culture (as are the cultural objects produced by any society) and that the resources of technology

[2] Genetics in everyday life: the use of a newsletter to disseminate and teach genetics. Available at: <
http://web2.sbg.org.br/congress/sbg2008/pdfs2008/EN.pdf>. Accessed on: Mar 2017

are part of scientific culture. In addition, the constant transformation of scientific knowledge and technology also makes it essential to develop in students the attitude that they will continue to learn throughout their lives.

Santos (2007) argues that science encompasses different social actors and that understanding this field depends on analyzing the interrelationships between them. Therefore, it can be considered that understanding the purposes of science education involves an analysis of the different ends that have been attributed to it by its various actors.

Based on this concept, the PCNs recommend that teachers bring information on current issues into the classroom, such as the Human Genome Project, the cloning of higher organisms and GMOs (BRASIL, 1997). These should be discussed in a dynamic and critical way, so that the student can reflect, because understanding genetic phenomena is not simple, as it involves invisible processes and entities that are not part of the students' day-to-day experiences (BRASIL, 1997).

In this sense, the teaching of Science and Biology should guide the student's position on these issues, as well as others, such as their day-to-day actions: caring for their bodies, eating and sexuality (BRASIL, 1997).

According to Brasil (2006), school knowledge should be structured in such a way as to make it possible to master the scientific knowledge systematized in formal education, recognizing its relationship with everyday life and the possibilities of using the knowledge learned in different situations in life. But for this to happen, the proposal depends on the teacher becoming a mediator between systematized knowledge and the student, so that the latter can transpose the content learned in the classroom into everyday life (BRASIL, 2006).

Carvalho (2004) argues that a teaching proposal aimed at the critical formation of the student must be planned in such a way as to lead the students to construct their conceptual content, participating in the process of building knowledge and giving them the opportunity to learn to argue and exercise reason. Instead of providing them with definitive answers or imposing their own points of view by transmitting a closed view of science.

It's important to say that in order to work with students on investigative science activities, they need to learn to think in order to solve the problems generated by the investigation. It is therefore considered that the knowledge socialized in the classroom is an instrument for mobilizing students to solve problems, where they can, based on their needs, produce their own hypotheses, thus generating knowledge through the interaction between thinking, feeling and doing, as emphasized by Azevedo (2004).

An investigative activity should start from a problematic situation and should lead students to reflect, discuss, explain, report, in short, to start producing their own knowledge through the interaction between thinking, feeling and doing (AZEVEDO, 2004). From this perspective, learning procedures and attitudes becomes just as important as learning concepts and/or content (CARVALHO, 2004).

This is why it is considered essential for teachers to work with this practice in the classroom. However, before carrying out the activities, the teacher needs to plan ahead and select the activities and teaching procedures that will be used, in order to promote conceptual development in the students towards scientifically accepted ideas (CARVALHO, 2004).

Silva and Pereira (2016) argue that in order to obtain satisfactory results in science teaching, it is necessary to rethink pedagogical practice. Therefore, according to them, it is up to the teacher to research methodologies that adapt to the student's reality and from there to promote experimental activities that can stimulate and help the student to understand the concepts and understand science as a historical construction and practical knowledge.

These should awaken the student's curiosity and creativity, and make them capable of using scientific information and knowledge to understand the world around them and solve problems and questions that are posed to them. This view is shared by Pereira (2012); Pereira; Ribeiro; Freitas (2014; 2016); Vinhas and Pereira (2016).

The literature also warns that the teacher should use experimental activities as an important resource for formulating questions about concrete reality, making predictions, testing the hypotheses raised, debating ideas and developing the student's capacity for argument, a critical and investigative stance, and finally for the student to be able to intervene in the environment in which they live (SILVA; PEREIRA, 2016).

5 MATERIALS AND METHODS

This is a qualitative-quantitative study. Minayo (2010, p. 54) points out that "the method has a fundamental function: to make it plausible to approach reality based on questions asked by the researcher".

This study involved 30 students from the 3rd year of secondary school at Rodoval da Silva Borges State School and 30 students from Barroso Tostes State School, both located in the city of Santana (AP). The only selection criterion used was that the interviewees were in the 3rd year of secondary school.

To collect the data, a questionnaire was used, containing open-ended questions aimed at investigating the students' level of knowledge and interest in transgenics, what they understand by biotechnology and the advantages and disadvantages of consuming TA. It should be noted that the questionnaire was administered on different days, due to the fact that there were two schools involved in the research and also because of the students' availability to answer it.

The questionnaire was completed during a period of class time (30 minutes) provided by the teacher for this purpose. At this point, the students were instructed on how to fill in the questionnaire and the questions were read out. The data was collected in December 2016.

This procedure was selected because, according to Fernandes (2007), it is a good way for researchers to obtain information about the opinions, attitudes and knowledge of the subjects who make up the study sample.

Data analysis used the Discourse of the Collective Subject - DSC, proposed by Lefreve and Lefreve (2006), which seeks to respond to the self-expression of collective thought or opinion, respecting the dual qualitative and quantitative condition of these as an object. The answers were categorized according to Bardin's (2011) proposal.

In this case, students were asked about the teaching of Science and Biology in two Santana public schools, and their responses served as the object of study, the results of which are presented in this research.

6 RESULTS AND DISCUSSIONS

The first question put to the students was: "*What are Transgenic Foods? Explain".* The data relating to this question was categorized. Of the 60 participants in the survey, 65% said that "they are genetically modified foods"; 20% said that "they are foods that have been altered in their original form"; 10% said that they are foods "with transgenics in their composition" and 5% said that they are "potatoes, salty snacks, corn, etc".

When analyzing the data, the majority (65%) said that transgenics are genetically modified foods. This shows that the interviewees have some knowledge of the subject, bearing in mind the concept developed by Lima (2005).

However, it should be noted that the term transgenic, by definition, refers to any organism, microorganism, animal or plant that, through transformation via genetic engineering, has had its genetic constitution altered by the introduction of gene(s) from another organism, generally from another species (PEDRANCINI et al., 2008; KREUZER; MASSEY, 2008; TORRES; CALDAS; BUZO, 1998).

On the other hand, the research subjects (100%) did not mention the use of genetic engineering when talking about this genetic modification, nor did they mention the use of genetic engineering or biotechnology.

This first question showed that most of the subjects in this study (65%) had a basic knowledge of what transgenics is. Pedrancini et al. (2008) found in their study that the knowledge taught at school on the subject has not allowed the subjects to understand the current reality, a fact corroborated in this study. It is therefore important that teacher training courses provide adequate scientific and technological literacy for future teachers.

Scientific literacy has various definitions, understandings and conceptions, according to different authors who discuss the subject. It is worth remembering that the aim here is not to delve deeper into the subject, but to highlight its importance for science teaching. The understanding of scientific literacy is based on the thinking of Miller (2000a, 2000b) and Chassot (2003; 2011), who understand it as the mastery of basic knowledge about science in order to make subjects capable of behaving as responsible consumers, and at the same time enabling these subjects to position themselves on issues relating to science policies, guaranteeing the effective participation of citizens in support for science.

Cachapuz et al. (2011) argue that the process of scientific literacy may promote an approach that considers problems from a broader point of view, assessing their possible repercussions. It is in this sense that we highlight the following:

> The lack of scientific literacy about transgenic processes acts as an **obstacle to a more conscious and independent position**. In this case, the student lacks a basic repertoire of biological knowledge to understand the possible effects caused by the techniques used and to correlate them with other genetic information [...] (SOUZA; FARIAS, 2011, p. 26-27, emphasis missing in original).

In these terms, we are referring to a concern already discussed by Pedrancini et al. (2008), that the knowledge taught at school has not allowed students to understand current reality, as well as to think, speak and act scientifically, which are necessary attitudes in scientifically literate people.

The following answers given by the students: "Alterations in their original form (20%), potatoes, corn, salty snacks (5%), with transgenics in the composition (10%)", indicate that they are not fully aware of the subject. At this point, it should be noted that the conceptual difficulties related to the subject of transgenics have already been pointed out in studies such as Silva and Calsa's (2003), which demonstrated some of the students' erroneous ideas about transgenics, including the fact that students understand that transgenic plants are artificial beings.

Pozo and Crespo (2009) believe that this confusion of ideas represents the way in which students often perceive scientific phenomena, i.e. in the wrong way. In this way, it is believed that a perspective of misinterpretation about the nature of science ends up, in a way, functioning as an epistemological barrier, "[...] revealing value judgments based on a ready-made, entrenched science, since students do not seem to depend on what they have learned to position themselves, since the function (good or bad) that science performs is defined" (SOUZA; FARIAS, 2011, p. 26).

In these terms, educational institutions, be they schools or universities, must provide the means for students to be able to recognize science as a human activity in constant transformation, resulting from the interaction of historical, social, political, economic, cultural, religious and technological factors, and therefore **not** neutral (BRASIL, 2006).

Like Chassot (2003, p. 31), we believe that "Our greatest responsibility in teaching science is to ensure that our students become more critical men and women as a result of our teaching". He continues: "We dream that, through our education, students can become agents of change - for the better - in the world in which we live" (CHASSOT, 2003, p. 31).

This is why the same author considers "[...] scientific literacy as a body of knowledge that would make it easier for men and women to read the world in which they live" (CHASSOT, 2003, p. 38). In fact, we need to set ourselves up as agents of change - teacher and pupil - in order to guarantee a better world.

The second question, *"What does the acronym GMO have to do with Transgenic Foods?"* had the students' answers organized into the following categories: 70% "did not answer"; 20% said it meant "Genetically Modified Organism"; 5% said it "indicates that the food is transgenic" and 5% said it was "the technical concept of genetically modified foods".

The majority (70% of students) did not answer the question. Those who answered "Genetically Modified Organism" were a minority (20%).

As highlighted elsewhere in the text, Lima (2005) states that GMOs are defined as any organism that undergoes human interference in its genome. Whether this is through simple genetic improvement (a technique which consists of crossing organisms until the best genome is selected) or through the direct manipulation of genes through genetic engineering. The author warns that it is a conceptual error to classify GMOs as transgenic, even though they are all GMOs. However, not all GMOs are transgenic (LIMA, 2005).

Considering that 70% of the 60 students who took part did not know the acronym in question, as can be seen from the data presented, the hypothesis is that the strike that took place during the year of the survey delayed the content.

Another hypothesis is that the teaching and learning process may be happening in a **mechanical** way, in which the student is only concerned with memorizing the formulas to take the tests, forgetting them immediately after their needs have been met. According to Ausubel (1980), this type of learning is characterized by an inability to use and construct knowledge.

On the other hand, what we want for effective learning is for it to be **meaningful** for the student. In science teaching, the "reinterpretation" of the concepts worked on in the classroom produces and expresses changes related to the student's cognitive structure. For this reason, the importance of the knowledge and skills that already exist in the cognitive structure of the learner should be emphasized. The literature points to cognitive structure as that formed by the content of ideas and their organization (AUSUBEL; NOVAK; HANESIAN, 1980). It is in this sense that **meaningful learning**

> [...] it is the process by which new information received by the subject interacts with a specific knowledge structure oriented by relevant concepts, the subsungor concepts - or incorporating, integrating, inserting, anchoring concepts - that determine the prior knowledge that anchors new learning (ALEGRO, 2008, p. 24).

In this respect, Moreira (1999, p. 13) points out that it is not simply a matter of association, but "[...] of interaction between specific and relevant aspects of the cognitive structure and new information, through which it acquires meaning and is integrated into the cognitive structure".

In this sense, Alegro (2008, p. 24) also points out that the sub-sung concepts "[...] are reworked, becoming more comprehensive and refined. Consequently, their meanings are improved and their potential for subsequent significant learning is enhanced". Learning meaningfully means, among other things, understanding the logical organization of what is to be learned.

Moreira (1999, p. 185) also characterizes meaningful learning as: "[...] the organization and integration of new material into the cognitive structure". For his part, Lemos (2006, p. 57) refers to meaningful learning as a product, because he characterizes it in the following terms: "[...] a meaning identified at a specific moment, however, is always a provisional product because at the next instant, depending on contextual factors and the subject's intentionality, this knowledge may change".

And this kind of observation can be very valuable for science teaching, since with constant advances, much of its knowledge - once considered true - is undergoing mutations.

Moreira (1999, p. 185) highlights another aspect of learning, namely machine learning, also called automatic learning or simple memorization. He says that it is "[...] a *continuum* and not a dichotomous opposition". In this respect, Alegro (2008, p. 25) believes that this type of learning "[...] is conceived as learning new information with little or no association with relevant concepts existing in the cognitive structure".

However, despite being criticized, authors such as Pontes Neto (2001, p. 65) point out that machine learning, or a "[...] certain degree of mechanicalness, should not be disregarded because content that cannot be substantively modified is also necessary on a daily basis".

In the teaching and learning process, especially in science, we must try to get closer to the students' reality, looking for ways to understand their universe and how they understand science, or at least the aspects related to it. Since our aim is to make the ideas presented by our actions as educators meaningful to learners.

Considering the fact that biotechnology subjects should be covered from the 2nd year of high school onwards, question 03 of the questionnaire sought to ascertain the students' conception of *"What is Biotechnology?".*

Of the 60 students interviewed, the following categories were drawn up: 40% "No answer"; 30%; "Part of Biology that studies genetic material"; 5% "Technology related to life"; 5% "Living organisms for the production of goods and services"; 10% "Second Generation Technology"; 5% "advanced technology"; 5% "studies genetic engineering".

The data shows that the majority (40% of interviewees) are also unable to define biotechnology. The rest, even though they made different hypotheses about what biotechnology is, were unable to correctly define the term.

According to Gongalves (2010), Biotechnology is a field of knowledge of a multidisciplinary nature that encompasses varied knowledge with the main emphasis on Chemistry, Biology and Microbiology and Molecular Genetics. At the beginning of the theoretical discussions in this work, Durbano et al. (2008) stated that Biotechnology is a set of techniques that have made it possible to use organisms to obtain products that are useful to human beings.

If we revisit texts from the 1980s, the years in which the term "biotechnology" spread, we find more than a dozen different definitions of the term. The following were among the most frequently found definitions:

- OECD - *Organization for Economic Co-Operation and Development:* The application of the principles of science and engineering to the treatment of materials by biological agents in the production of goods and services (1982).
- OTA - *Office of Technology Assessment:* Biotechnology, broadly speaking, includes any technique that uses living organisms (or parts thereof) to obtain or modify products, improve plants and

animals, or develop microorganisms for specific uses (1984).

- EFB - *European Federation of Biotechnology:* Integrated use of biochemistry, microbiology and engineering to apply the capabilities of microorganisms, cultured animal or plant cells or parts thereof to industry, health and environmental processes (1988).
- E.H. Houwink: the controlled use of biological information (1989).
- *Biotechnology Industry Organization:* in a broad sense, Biotechnology is "bio" + "technology", i.e. the use of biological processes to solve problems or make useful products (2003).

It can be seen that, over time, the concept has gained a simpler expression. The most recent definitions no longer refer to the technological processes involved; perhaps because, as well as being complex and diverse, these are evolving very quickly.

In this text, biotechnology is considered broadly in the terms of Malajovich (2009), defined as an activity based on multidisciplinary knowledge, which uses biological agents to make useful products or solve problems. This definition is broad enough to encompass activities as varied as those of engineers, chemists, agronomists, veterinarians, microbiologists, biologists, doctors, lawyers, businesspeople, economists etc.

Since the students were unable to refer to the definition adopted here, the assumption is that, although the teachers worked on the concept of Biotechnology with the students interviewed, it may be that the learning was not significant and occurred mechanically, being forgotten as soon as the student had met their need, which was probably one of the bimonthly exams.

Considering the importance of learning being
and that is capable of transforming the student's social reality, it was important to find out, through the fourth question of the questionnaire *"For you, what are the advantages of consuming food?*
transgenic?". Regarding this question, the data was organized into the following categories: 70% "no answer"; 10% "no agrotoxins"; 10% "food can be enriched".
nutritionally"; 5% "tastier" and 5% "prevents food pests".

Looking at the data, it can be seen that the majority (70%) did not answer the question, i.e. they have no knowledge of the advantages of consuming such foods. The answers related to the non-use of agrotoxins and the absence of pests have to do with production, not the consumption of TA.

Ometto and Toledo (2003) state that GMOs were developed and marketed because they offer advantages to producers and/or consumers, such as lower production costs, lower prices, greater nutritional value and product durability. They also point out that there is no evidence that new proteins present in genetically modified foods are more allergenic than traditional proteins.

In this case, the students' lack of knowledge about TA and, consequently, biotechnology is also evident. This result was to be expected, given the reality of public schools in the state of Amapa. Most of them don't have laboratories or resources that can be used by teachers to encourage students to do research, so teaching is stuck with the textbook, which is meaningless and discouraging for students.

In fact, the literature points out that "Transgenic crops will manifest the following types of 'desired characteristics': [...] more nutritious grains; [...] 'intensifications' of the food's aesthetic appeal (taste, texture, appearance) or market demands" (LACEY, 2006, p. 99).

These characteristics conferred on transgenics are also pointed out by Green (2008, p. 17): "Genetic modifications can produce sweeter strawberries"; "The science of genetic manipulation can create juicier tomatoes" (GREEN, 2008, p. 7); "Transgenic corn has large, juicy seeds" and "The transgenic wheat produced today has larger grains" (GREEN, 2008, p. 10); "Scientists are using transgenics to make animals grow bigger and faster [...] genes are added to produce animals with less fat and more meat" (GREEN, 2008, p. 21).

Thus, the literature shows a wide range of improvements in the quality of food derived from transgenics, corroborating the justification given by a small proportion of the interviewees.

Of the interviewees who responded to the questionnaire and the few answers obtained, among them: *"Prevents pests in food (5%); no agrotoxins (10%)",* it was observed that, corroborating some of the students' statements, Almeida and Lamounier (2005, p. 351) state that transgenic plants "[...] offer the possibility not only of bringing desirable characteristics from other varieties of the plant, but also of adding characteristics from other unrelated species". They go on to say that "Transgenics also have other important characteristics, such as **resistance to pests and herbicides** [...]" (ALMEIDA; LAMOUNIER, 2005, p. 351 - emphasis missing in original).

With regard to resistance to pests and a reduction in the use of agrotoxics, Valicente and Andrade (2011, p. 215 - emphasis missing in the original) point out that:

> The use of genetically modified plants has led to a **reduction in**

> **the use of insecticides**, especially broad-spectrum ones, **which favors** the maintenance of natural enemy populations and their adaptation to **pest control**. Therefore, transgenic plants are not a substitute for insecticides, but rather a new approach within IPM [Integrated Pest Management], as they eliminate or drastically reduce the use of non-selective insecticides, favoring the alliance with biological control.

Borem and Carneiro (2008, p. 405) state that "Genetic transformation has the potential to improve productivity, resistance, nutritional quality and other characteristics of cultivated plants". For their part, Siqueira and Trannin (2008, p. 239) state that "The rapid adoption of transgenic crops around the world is a consequence of the benefits brought by this technology to farmers, consumers and society in both industrial and developing countries".

Proponents of transgenics argue for the legitimacy and importance of the development, immediate implementation, intensive use and widespread diffusion of transgenics: "[...] in the agricultural practices that produce the largest harvests around the world as quickly as possible, and thereby become a central platform in public agricultural policies" (LACEY, 2006, p. 32).

These groups rely on "[...] the prestige of technoscience[3] , maintaining that transgenic technology is effective, beneficial, legitimate and even has a mandatory place in national agriculture and trade policies" (LACEY, 2006, p. 34). They also claim that transgenics "[...] are examples of technoscientific developments, which are the main origin of the improvement of agricultural practices and the satisfaction of human needs" (LACEY, 2006, p. 35).

---

[3] Technoscience refers to science conducted with the aim of technological development and to technological advances that are informed by science conducted according to materialist strategies.

The author also points out that these groups claim that there are no alternative forms of agriculture that can be developed to replace transgenics without posing unacceptable risks to humanity, such as not producing enough food to feed and nourish the world's growing population. Furthermore, proponents assure that "[...] they could reasonably be expected to lead to major benefits in terms of productivity, sustainability and meeting human needs" (LACEY, 2006, p. 37). It's that old, but still widely held argument that "transgenics are necessary to feed the world" (ZANONI; FERMENT, 2011; OLIVEIRA, 2004).

For Lacey (2006, p. 29), transgenics "[...] are rapidly taking shape and largely bringing about changes in agricultural practices in various parts of the world and, at the same time, facing strong resistance from various interest groups". In fact, the issue of transgenic crops has been much debated (BOREM; CARNEIRO, 2008; CARPENTIERI-PIPOLO, 2009; CARPENTIERI-PIPOLO; MARASCHIN, 2009; CERDEIRA et al., 2009; CRAIG; DEGRASSI; RIPANDELLI, 2009; LACEY, 2006; MARCELINO et al, 2009; ODA; SOUZA; BORGES, 2009; PATERNIANI, 2009; PEREIRA; VIEIRA, 2009; VALICENTE, 2009), including in the field of science teaching (SOUZA; FARIAS, 2011; PEDRANCINI et al., 2007; 2008; DURBANO et al.., 2008; FABRICIO et al., 2006; BOSSOLAN, 2008; SANTOS; MARTINS, 2009; TAKAHASHI; MARTINS; QUADROS, 2008; PEREIRA, 2012; PEREIRA, RIBEIRO; FREITAS, 2014; 2016; VINHAS; PEREIRA, 2016).

At this point, it should be emphasized that a more comprehensive discussion on this topic would not only help students to position themselves for or against the new technologies coming from this area, but would also enable students to reach the "[...] multidimensional stage of scientific literacy, having a more integrated understanding of the concepts and processes learned and establishing relationships between this knowledge and science, technology and society" (CAMARGO; INFANTE-MALACHIAS; AMABIS, 2007, p. 11).

According to this perspective, we are contributing to the formation of scientifically literate students, capable of making their own decisions based on what is best not only for them, but for the other citizens of the world, whose greatest asset is life in the only known sphere capable of offering conditions of existence for our species and for others (PEREIRA, 2012).

Once the students' knowledge of the advantages of consuming *transgenic foods had* been verified, it was considered important to find out what they thought of the disadvantages, so this question was asked in the fifth question of the survey: *"What do you think are the disadvantages of consuming transgenic foods?".* The data was organized into the following categories: 30% "unreliable"; 20% "may cause damage to the environment"; 20% "no answer"; 15% "may cause damage to health"; 10% "may cause unexpected results to the human organism" and 5% "not natural products".

In this case, the data shows a very wide variety of answers. It can be seen that most of the answers were based on empirical knowledge, common sense, without actually bringing scientific knowledge to the students' speeches.

According to Nodari and Guerra (2000), among the probable risks generated by the consumption of TA is the case of absorption of Deoxyribonucleic Acid (DNA) by eukaryotic cells and a second type of risk is related to adverse reactions to GMO-derived foods, which, according to their effects, can be classified into two groups, namely: allergenic and intolerant. Allergenic foods cause allergic hypersensitivity.

The second group is responsible for physiological changes, such as abnormal or idiosyncratic metabolic reactions and toxicity. There are also a number of other risks to human health that must be analyzed using appropriate protocols (NODARI; GUERRA, 2000).

Groups opposed to TA claim that the argument about the risks related to this type of food is not scientifically well established (PEREIRA; RIBEIRO; FREITAS, 2016). Furthermore, the greatest risks may not be directly related to human health and the environment, mediated by biological mechanisms, but may be those caused by the socio-economic context of transgenic research and development and its associated mechanisms (PEREIRA, 2012). These mechanisms include "[...] the stipulation that transgenic seeds are objects in relation to which intellectual property rights must be guaranteed" (LACEY, 2006, p. 37).

It is for these and other reasons that transgenics are much debated and divide opinion (PEREIRA; RIBEIRO; FREITAS, 2016). This is because while studies point to the benefits of using transgenics, others do not. In addition, many authors only point out its negative aspects, without trying to "relativize" and understand what they are and what their possible benefits are (PEREIRA; RIBEIRO; FREITAS, 2016).

The author referred to below talks about the potential risks and benefits of TA. For her, these risks and/or benefits

> [...] must be taken into account **in decision-making** when drawing up strategies for standards and policies which, while [sic] promoting development, protect life, human and animal health and the environment (CARPENTIERI-PIPOLO, 2009, p. 25).

For this reason, it is understood that in order to make a certain decision, the citizen needs to have information and the critical ability to evaluate it in order to make the most informed choices, taking into account the costs and benefits of the decision. Thus, qualified decision-making is necessary, but for this it is essential to have qualified scientific knowledge, which will only be possible once the student is scientifically literate (PEREIRA; RIBEIRO; FREITAS, 2016).

Of the 15% who answered that: "They can cause damage to health" is in line with Lima's (2001) argument that the health problems involved with TA are obvious. According to the author, many guinea pigs fed these foods and tested in laboratories showed allergic reactions. In addition, the person or animal ingests food that has been genetically transformed to resist pesticides, so in this case, it can be concluded that the individual runs the risk of being fed poison residues. In the same vein, "The transfer of genes from Brazil nuts to soybeans produced allergic reactions in people" (NORDLEE et al., 1996).

There will certainly still be a lot of debate about the advantages and disadvantages of consuming transgenic foods, as there are those who defend and those who are against their production and consumption. This dispute is reflected in the production of these foods, because in some countries they are banned precisely because they have not been approved by the bodies responsible for their release.

FINAL CONSIDERATIONS

After studying this research and analyzing the data, it can be said that students in the 3rd year of secondary school in public schools in the municipality of Santana-AP did not have in-depth knowledge of biotechnology, especially TA. It appears that this is possibly a consequence of the mechanized way in which the content is taught by the teachers.

It is argued here that subjects such as biotechnology and TA should be discussed more emphatically in the classroom, using resources that help students to problematize, hypothesize, interpret and understand what is being taught in a meaningful way. It is believed that this would be one of the possible ways in which science learning actually takes place.

The teacher needs to work in a dynamic way so that the student feels motivated to take part in the lessons. It is therefore considered that in order to obtain satisfactory results in the teaching of biotechnology, it is essential for the teacher to use methodologies that encourage the student's contact with the object of study, so that they can observe its characteristics, the natural phenomena that occur and its many facets. In short, they need to be able to reflect,
question, investigate, and get involved with the activity by applying your knowledge.

Finally, it is important not to lose sight of the need to address current issues such as TA in the classroom. It's a controversial subject that divides the opinions of governments, nations, the population and even scientists. It is essential that people are equipped to make qualified decisions, both collectively and individually. Furthermore, preparing people to make decisions is an educational objective that is of fundamental importance for a citizen's education, capable of responding to the challenges of today's society.

This task is also a challenge for current and future teachers, especially when it comes to identifying the best approach to socio-technical issues. Certainly, the discussions must reflect the needs of societies, now and in the future. It is in these terms that the role of scientific literacy in science teaching is reinforced here, with a view to guaranteeing subjects' autonomy and a critical stance, which is fundamental in the constitution of conscientious citizens.

REFERENCES

ALEGRO, Regina Celia. **Prior knowledge and significant learning of historical concepts in secondary education**. 239f. Thesis (Doctorate in Education). Universidade Estadual Paulista Julio de Mesquita Filho, Marilia. 2008.

ALMEIDA, Gustavo Calixto Scoralick de; LAMOUNIER, Wagner Moura. Transgenic foods in Brazilian agriculture: evolution and prospects. **Organizagoes Rurais & Agroindustriais**, Lavras, v. 7, n. 3, p. 345-355, 2005.

ARAGAO, Francisco Jose Lima. **Transgenic organisms**. Barueri: Manole, 2002. 115 p.

ARRUDA, Ana Maria da Silva; BRANQUINHO, Fatima Teresa Braga; BUENO, Shirley Neves. **Science in primary education**. January, 2006. Available at: <http://docplayer.com.br/1919243-Ciencias-no- ensino-fundamental.html>. Accessed on: January 12, 2017.

AUSUBEL, David Paul. **Acquisition and retention of knowledge**: a cognitive perspective. Lisbon: Platano, 2003. 219 p.

AUSUBEL, David Paul; NOVAK, Joseph D.; HANESIAN, Helen. **Educational Psychology**. Rio de Janeiro: Interamericana, 1980. 626 p.

AZEVEDO, Maria Cristina P. Stella de. Teaching by investigation: problematizing classroom activities. In: CARVALHO, Anna Maria Pessoa de. (Org.). **Ensino** de **ciencias**: unindo a pesquisa e a pratica. Sao Paulo: Pioneira Thomson Learning, 2004.

BARDIN, Lawrence. **Content Analysis**. Sao Paulo: Edigoes 70, 2011. 279 p.

BOREM, Aluizio; CARNEIRO, Jose Eustaquio de Souza. Gene flow. In: BOREM, Aluizio; GIUDICE, Marcos Del (Org.). **Biotechnology and the Environment**. Vigosa: Federal University of Vigosa, 2008. p. 405-419.

BOSSOLAN, Nelma Regina Segnini. The theme of biotechnology in the teacher's manual: additional readings and activities for teacher training. In: PAVAO, Antonio Carlos; FREITAS, Denise de. (Org.). **Quanta ciencia**

ha no ensino de **ciencias.** Sao Carlos: Ed. UFsCar. 2008. p. 301-308.

BRAZIL. Ministry of Education (MEC). Secretariat for Basic Education. **Orientações Curriculares para o Ensino Medio**: Ciencias da Natureza, Matematica e suas Tecnologias, v. 2. Brasilia: MEC. 135 p, 2006.

CACHAPUZ, Antonio et al. (Org.). **The necessary renewal of science teaching**. Sao Paulo: Cortez, 2011. 263 p.

CAMARGO, Solange Soares; INFANTE-MALACHIAS, Maria Elena; AMABIS, Jose Mariano. The teaching of molecular biology in colleges and high schools in Sao Paulo. **Revista Brasileira do Ensino de Bioquimica e Biologia Molecular**. v. 5, n. 1, p.1-14, 2007.

CARPENTIERI-PIPOLO, Valeria. Regulation of genetically modified organisms: challenges and opportunities for the biotechnology. In: CARPENTIERI-PIPOLO, Valeria (Org). **Transgenic crops**: an approach to benefits and risks. Londrina: EDUEL, 2009. p. 301-320.

CARPENTIERI-PIPOLO, Valeria; MARASCHIN, Marcelo. Transgenic food safety: new food possibilities in the detection of GMOs in food. **Transgenic crops**: an approach to benefits and risks. Londrina: EDUEL, 2009. p. 229264.

CARVALHO, Ana Maria Pessoa de. Structuring criteria for science teaching. In: CARVALHO, Ana Maria Pessoa de (Org.). **Teaching Science**: linking Research and Practice. Sao Paulo: Pioneira Thomson Learning, 2004. p. 1-17.

CERDEIRA, Antonio et al. Herbicide-resistant transgenic plants and interactions with the environment. In: CARPENTIERI-PIPOLO, Valeria (org). **Transgenic crops**: an approach to benefits and risks. Londrina: EDUEL, 2009. p. 153-172.

CHASSOT, Attico. **Scientific Literacy**: issues and challenges for education. 3 ed. Ijui: Unijui, 2003. 440p.

CHASSOT, Attico. **Scientific literacy**: issues and challenges for education. 6 ed. Ijuf: Unijrn, 2014. 368p.

COLL, Cesar. **School learning and the construction of knowledge**.

Porto Alegre: Artes Medicas, 1995. 159 p.

CRAIG, Wendy; DEGRASSI, Giuliano; RIPANDELLI, Decio. Towards the safe use of modern biotechnology: an assessment of the potential negative impacts of transgenic crops and their products and derivatives. In: CARPENTIERI-PIPOLO, Valeria (org). **Crops Transgenics**: an approach to benefits and risks. Londrina: EDUEL, 2009. p. 27-66.

DARWIN, Charles. **The Origin of Species by Means of Natural Selection or the Preservation of Favoured Races in the Struggle for Life**. New York: Modern Library, 1859. 549 p.

DURBANO, Joao Paulo De Monaco et al. Perception of the knowledge of high school students in the municipality of Joao Pessoa-PB on emerging issues in biotechnology. In: **Anais...** BRAZILIAN CONGRESS OF GENETICS, LIV Salvador, 2008.

FABRICIO, Maria de Fatima Lima et al. The understanding of Mendel's Laws by Biology students in Basic Secondary Education. **Ensaio - Pesquisa em Educagao em Ciencias**, v. 8, n. 1, p. 1-21,2006.

FERNANDES, Carla Alberta da Fonte. **Mathematics in the subject of Physical and Chemical Sciences: a study on the attitudes of 9th grade students**. Dissertation [Master's in Education]. Braga: University of Minho. 2007.

FURNIVAL, Ariadne Cloe; PINHEIRO, Sonia Maria. Public perception of information on the potential risks of transgenics in the food chain. **Historia, Ciencias, Saude - Manguinhos**, Rio de Janeiro, v. 15, n. 2, abr.-jun., p.277-291,2008.

GONQALVES, Eric Armendani. **Biotechnology: Social representations of chemistry teachers**. 121 f. Dissertation (Master's in Biotechnology). University of Mogi das Cruzes. Mogi das Cruzes-SP, 2010.

GORGEN, Sergio Antonio (Org.) **Riscos dos transgenicos.** Rio de Janeiro: Ed. Vozes, 2000. 92 p. (Biodiversity collection & transgenics).

GREEN, Jen. **Transgenic Foods**. Sao Paulo: DCL, 2008. 31p.

GUERRANTE, Rafaela Di Sabato. **Transgenics**: a vision estrategica. Rio de Janeiro: Interciencia. 2003. 173 p.

KREUZER, Helen; MASSEY, Adrianne. **Recombinant DNA and biotechnology**: a guide for teachers. 8. ed. Washington: ASM Press, 2008. 704 p.

LACEY, Hugh. **The transgenic controversy**: scientific and ethical issues. 1 ed. Aparecida, Sao Paulo: Ideias e Letras, 2006. 239 p.

LEMOS, Evelyse dos Santos. Meaningful Learning: facilitating strategies and evaluation. In: **Dossie do I Encontro Nacional de Aprendizagem Significativa**. Serie Estudos, UCDB, n. 21, p. 53-66, jun./2006. Campo Grande-MS.

LEVREFE, F. LEVREFE, A. M. C. O sujeito coletivo que fala. **Interface - Comunic, Saude, Educ,** v. 10, n. 20, p. 517-24, jul./dez., 2006.
LIMA, L. M. **Transgenicos**: da Origem a Polemica. Rio de Janeiro, 2005.

LOURENQO, A.P.; REIS, L. G. Transgenics in the classroom: conceptions and opinions of high school students and a pedagogical practice. **Revista Vozes dos Vales da UFVJM**: Academic Publications - MG - Brazil, n. 3, Year II, 2013.

MAIA, L. C. Alimentos geneticamente modificados e o Codigo de Defesa do Consumidor. **Ambito Juridico**, Rio Grande, XIV, n. 93, Oct., 2011.

MALAJOVICH, Maria Antonia. **Biotechnology Fundamentals**. Rio de Janeiro: ORT, 2009. 104 p.

MARCELINO, Francismar Correa et al. Detection and quantification of transgenic grains and foods: the Brazilian panorama. In: CARPENTIERI-PIPOLO, Valeria (Org.). **Transgenic crops**: an approach to benefits and risks. Londrina: EDUEL, 2009. p. 265-288.

MILLER, Jon D. Scientific literacy and citizenship in the 21st[s] century. In: SCHIELE, B.; KOSTER, E. (Org). **Science centers for this century**. Quebec: Multimondes, 2000b. p. 369-411.

MILLER, Jon D. Scientific literacy: a conceptual and empirical review. **Daedalus**, v. 112, n. 2, p. 29-48, 1983.

MILLER, Jon D. The development of civic scientific literacy in the United States. In: KUMAR, David; CHUBIN Daryl E. (Org.), **Science, technology and society**: a sourcebook on research and practice. New York: Kluwer Academy/Plenum. 2000a. p. 21-47.

MINAYO, Maria Cecilia de Souza. **The challenge of knowledge**: qualitative research in health. 12 ed. Sao Paulo: Hucitec, 2010. 407p.

MOREIRA, Marco Antonio. Ausubel's theory. In: **Significant Learning**. Brasilia: Editora UnB, 1999. 129 p.

NODARI, Rubens Onofre; GUERRA, Miguel Pedro. Implications of transgenics for environmental and agricultural sustainability. **Historia, Ciencias, Saude - Manguinhos**, Rio de Janeiro, v.7, n.2, p.481-491, 2000.

NORDLEE, Julie A. et al. Identification of a Brazil-Nut Allergen in Transgenic Soybeans. **The New England Journal of Medicine**, v. 334, n. 11, p. 688-692, 1996.

ODA, Leila Macedo; SOUZA, Lucia de; BORGES, Kelly C. Almeida de Souza. Public perception and its impact on the regulation and development of agricultural biotechnology. In: CARPENTIERI-PIPOLO, Valeria (Org.). **Transgenic crops**: an approach to benefits and risks. Londrina: EDUEL, 2009. p. 355-378.

OLIVEIRA, Rui. Obstacles to science communication: the case of genetically modified organisms. **Communication and society**. v. 6, 2004.

OMETTO, Vanessa de Souza Rinaldo; TOLEDO, Simone Seghese de. Transgenics and EMBRAPA. **Revista de NutriQao**. jan./mar. 2003, vol.16, no 1, p.106.

PATERNIANI, Maria Lidia Stip. Transgenic crops and biodiversity conservation. In: CARPENTIERI-PIPOLO, Valeria (Org.). **Transgenic crops**: an approach to benefits and risks. Londrina: EDUEL, 2009. p. 131-152.

PEDRANCINI, Vanessa Daiana et al. Biology teaching and learning in high school and the appropriation of scientific and biotechnological knowledge. **Revista Electronica de Ensenanza de las Ciencias**. v. 6, n.

2, p. 299-309, 2007.

PEDRANCINI, Vanessa Daiana et al. Scientific knowledge and spontaneous knowledge: high school students' opinions on transgenics. **Ciencia & Educagao**, v. 14, n. 1, p. 135-146, 2008.

PEREIRA, Gerlany de Fatima dos Santos. **Appropriation of Scientific Knowledge: an approach to Food Transgenics**. 120 f. Dissertation (Master's Degree in Science and Mathematics Education). Postgraduate Program in Science and Mathematics Education (PPGECM), Institute of Mathematics and Science Education (IEMCI), Federal University of Para (UFPA), 2012.

PEREIRA, Gerlany de Fatima dos Santos; RIBEIRO, Elinete Oliveira Raposo; FREITAS, Nadia Magalhaes da Silva. **Appropriation of Scientific Knowledge**: an approach to Food Transgenics. 1. ed. Saarbrucken, Germany: Verlag-Novas Edigoes Academicas, 2014. v. 1. 96p.

PEREIRA, Gerlany de Fatima dos Santos; RIBEIRO, Elinete Oliveira Raposo; FREITAS, Nadia Magalhaes da Silva. **Socio-scientific controversies in science teaching**: AGROBIO in focus. 1. ed. Saarbrucken, Germany: Verlag-Novas Edigoes Academicas, 2016. v. 1. 94p.

PEREIRA, Luiz Filipe Protasio; VIEIRA, Luiz Gonzaga Esteves. Biosafety and new strategies for genetic transformation. In: CARPENTIERI-PIPOLO, Valeria (Org.). **Transgenic crops**: an approach to benefits and risks. Londrina: EDUEL, 2009. p. 213-228.

PONTES NETO, Jose. A. da S. On meaningful learning at school. In: MARTINS, Edna Julia S. et. al. **Different faces of education**. Sao Paulo: Arte & Ciencia Villipress, 2001, p. 13-37.

POZO, Juan Ignacio; CRESPO, Miguel Angel Gomez. **Learning and teaching science**: from everyday knowledge to scientific knowledge. Porto Alegre: Artmed. 2009. 296 p.

SANTOS, Wildson Luiz Pereira dos. Scientific education from the perspective of literacy as a social practice: functions, principles and challenges. **Revista Brasileira de Educagao** v. 12 n. 36, Sep./Dec. 2007.

SANTOS, Eunice; MARTINS, Isabel P. Teaching about genetically modified foods. Contributions to responsible citizenship. **Revista Electronica de Ensenanza de las Ciencias**. v. 8, n. 3, p. 834-858, 2009.

SILVA, Pedro Aurelio de Queiroz Pereira da. Competition law and the regulation of public services. **Ciencia Juridica**, v. 19, n.123, p.13, May/June 2005.

SILVA, Rafael Gustavo Rigolon da; CALSA, Geiva Carolina. A study of high school students' concepts of transgenic plants. In: XI Pedagogy Week: teacher training for the 21st century. **Proceedings...** Maringa: State University of Maringa. 2003.

SILVA, Jessica Alves da; PEREIRA, Gerlany de Fatima dos Santos. **Experimentation in the process of teaching and learning science**: possibilities and limitations. 1. ed. Saarbrucken, Germany: Verlag-Novas Edigoes Academicas, 2016. v. 1.72p.

SIQUEIRA, Jose Osvaldo; TRANNIN, Isabel Cristina B. Transgenic agrosystems. In: BOREM, Aluizio; DEL GIUDICE, Marcos (Orgs). **Biotechnology and the Environment**. Vigosa: Federal University of Vigosa, 2008. p. 225-310.

SOUZA, Aline Furtuozo de; FARIAS. Gilmar Beserra de. Perception of high school students' knowledge of transgenics: conceptions that influence decision-making. **Experiencias em Ensino de Ciencias**. v. 6, n.1, p. 21-32, 2011.

TAKAHASHI, Jacqueline Aparecida; MARTINS, Polyana Fabricia Fernandes; QUADROS, Ana Luiza de. Technological issues permeating chemistry teaching: the case of transgenics. **Quimica nova na escola**. n. 29, p. 3-7, aug., 2008.

TORRES, Antonio Carlos; CALDAS, Linda Styer; BUZO, Jose Amauri. **Tissue culture and genetic transformation of plants**. Brasilia: Embrapa, 1998. 509 p.

VALICENTE, Fernando Hercos; ANDRADE, Gilberto Santos. Genetically Modified Plants and Elements of Agricultural Entomofauna. In: BOREM, Alwzio; ALMEIDA, Gustavo Dias de (eds). **Genetically Modified Plants**: Challenges and opportunities for tropical regions. Visconde de Rio

Branco: Suprema, 2011. p. 205-217.

VALICENTE, Fernando Hercos. Risk assessment of insect-resistant transgenic crops. In: CARPENTIERI-PIPOLO, Valeria (org). **Transgenic crops**: an approach to benefits and risks. Londrina: EDUEL, 2009. p. 173-192.

VASCONCELOS, Maria Jose Vilaga de; CARNEIRO, Andrea Almeida; VALICIENTE, Fernando Hercos. Case study on Bt corn. In: BOREM, Alwzio; ALMEIDA, Gustavo Dias de (eds). **Genetically Modified Plants**: Challenges and opportunities for tropical regions. Visconde de Rio Branco: Suprema, 2011. p. 311-332.

VINHAS, Beatriz Fortes; PEREIRA, Gerlany de Fatima dos Santos. **Transgenic foods in the classroom**: considerations for teaching natural sciences. 1. ed. Saarbrucken, Germany: Verlag-Novas Edigoes Academicas, 2016. v. 1.67 p.

WEID, Jean Marc Von Der. Is the WHO "embracing" transgenics? **Radis**: comunicagao em saude, Rio de Janeiro, n.36, p.19, aug. 2005.

ZANONI, Magda; FERMENT, Gilles (Orgs). **Transgenics for whom**? Agriculture, Science and Society. Ministry of Agrarian Development. Brasília, 2011.538 p.

Printed by Books on Demand GmbH, Norderstedt / Germany